# Mushrooms, Molds and
# BREAD BUGS!

Rebecca Woodbury, Ph.D., M.Ed.

**Gravitas Publications Inc.**

# Mushrooms, Molds, and
# BREAD BUGS!

Illustrations: Janet Moneymaker

Mushrooms, Molds, and Bread Bugs!
ISBN 978-1-950415-54-0

Published by Gravitas Publications Inc.
Imprint: Real Science-4-Kids
www.gravitaspublications.com
www.realscience4kids.com

RS4K

Photo credits: Cover and Title Page: By Stephan Morris, AdobeStock; Above, By Andreas from Pixabay; P.3. By vivi Vivis from Pixabay; P.7. Top, Image by Andreas from Pixabay; Middle, ScottBauer/USDA/ARS; Bottom, CDC/DrLiberoAjello; P.9. By Tatiana, AdobeStock; P.11. By bela_zamsha, AdobeStock; P.14. By Michael Gäbler, CC BY SA 3.0; P.15. By Amadej Trnkoczy (amadej), CC BY SA 3.0; P.17. By Stephan Morris, AdobeStock; P.19. By yod67, AdobeStock; P.21. Public Domain

# Do you like **mushrooms**?

# Or **moldy** cheese?

I will try it!

Have you ever used **yeast**
to make yummy bread?

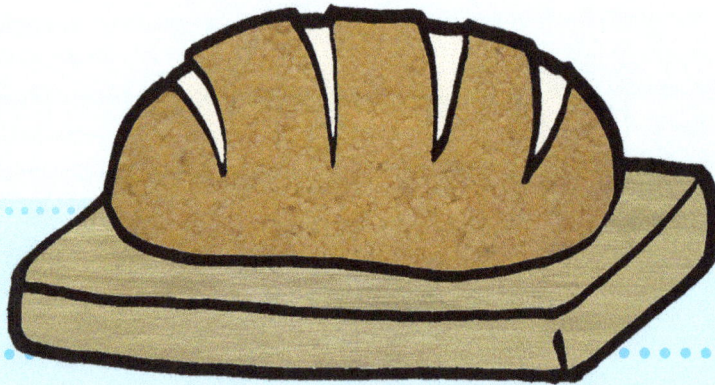

Yeast makes bread rise.

Yes! Yeast is what makes bread have air holes.

Mushrooms, molds, and yeast are called **fungi.**

Which one is this?

It is a mushroom.

**mushrooms**

**mold**

**yeast**

(seen with a
microscope)

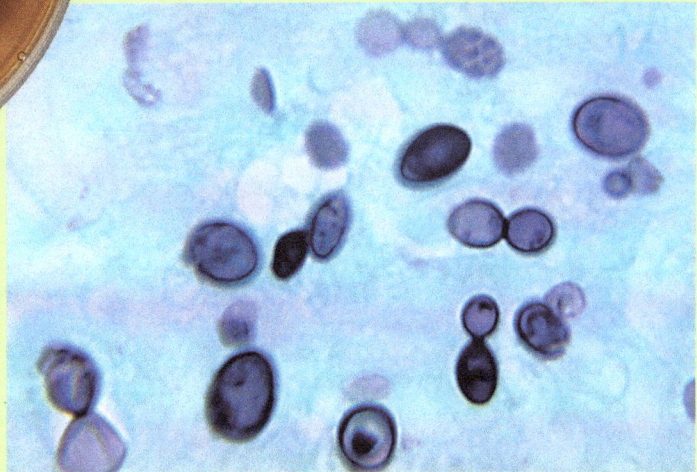

Fungi are found
almost everywhere.

In the dirt...

in the bathroom...

...in lakes, rivers, oceans,

and even in soup!

We can eat some fungi.

The cheese is the best part.

But some fungi are
poisonous and make us sick.

Some fungi are very BIG.

And some fungi are very small.

They are growing on little twigs!

# Fungi do not make their own food.

Do you have a shopping list?

Yes. Cheese and peanut butter.

- 16 -

Instead, they eat dead plants and animals.

Because fungi eat dead plants and animals, they are very important!

Fungi break down dead plants and animals into atoms and molecules.

Fungi leave atoms and molecules in the soil.

Plants use some of these atoms and molecules to make their own food and food for animals.

Fungi also make "fairy rings."
Legend says fairy rings are places
where fairies have danced.

But fairy rings are really mushrooms that are connected in the ground.

Sometimes the mushrooms grow in a circle. This circle is called a fairy ring.

We are making a fairy ring!

No fairies needed!

# How to say science words

**fungus**   (FUN-guhs) singular

**fungi**   (FUN-jiy) plural

**mushroom**   (MUSH-room)

**mold**   (MOHLD)

**yeast**   (YEEST)

**poisonous**   (POY-zuh-nuhs)

www.ingramcontent.com/pod-product-compliance
Lightning Source LLC
Chambersburg PA
CBHW040151200326
41520CB00028B/7563